The State Fair Book

Jack Pierce

The State Fair Book

Jack Pierce

Carolrhoda Books, Inc., Minneapolis

To my niece, Deb

Special thanks to Jerry Hammer, Publicity Superintendent for the
Minnesota State Fair.

LIBRARY OF CONGRESS CATALOGING IN PUBLICATION DATA

Pierce, Jack.
 The state fair book.

 SUMMARY: Photographs and brief text describe the various
attractions at a state fair including the animal and craft exhibits,
midway, and grandstand shows.

 1. Agricultural exhibitions—Pictorial works—Juvenile literature. 2. Fairs—Pictorial works—Juvenile literature. [1. Agricultural exhibitions—Pictorial works. 2. Fairs—Pictorial works]
I. Title.

S552.5.P53 1980   630'.74'01   79-91308
ISBN 0-87614-124-6   lib. bdg.

1   2   3   4   5   6   7   8   9   10   86   85   84   83   82   81   80

A Note from the Author

Fairs have been held all over the world for thousands and thousands of years. They began as a way of carrying on trade. Later contests and entertainments were added.

In 1810, the Berkshire Agricultural Society held a fair in Pittsfield, Massachusetts. It was called the Berkshire Cattle Show, and it was the beginning of state fairs in the United States as we know them today.

The photographs in this book were taken at the Minnesota State Fair. It's one of the biggest in the country, but it's very much like any other state fair. Companies exhibit their newest equipment and products. Dancers and blacksmiths show off their skills. Race-car drivers come to compete. Celebrities come to perform.

An important part of the fair is the competitions. 4-H members bring their horses, cows, pigs, and other livestock. Others bring pies, cakes, jams and jellies, and other foods. Still others bring items they have made during the year: quilts, sweaters, woodwork, and more. The judges award a blue ribbon for first place, red for second, white for third, and pink for fourth. They award ribbons for ten places in all. Then they judge all the first-place winners in a category and choose a grand champion. The grand champion receives a purple ribbon and the runner up gets a lavender ribbon.

State fair time is also entertainment time. Music, dancing, games, rides, circus and stage performers, a wide variety of foods—all create a festive mood.

I hope you'll have fun at the fair!

Horses

The English riders' jumping competition

Waiting for the judge to inspect your horse is part of the excitement. These horses are in the Western class.

The Black-smith

Metal is heated in the forge until it is soft.

Then the blacksmith hammers the hot metal into new shapes, like this plant hanger.

Horseshoeing

Horseshoes are heated and bent to fit the horse's hoof. Horses' hooves, just like people's feet, are different sizes and shapes.

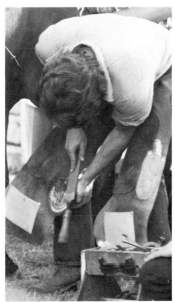

The horse's hoof is prepared for shoeing by filing it smooth.

Aerial Thrill Circus

These trapeze acrobats are three brothers.
They travel from fair to fair with their father,
who is also part of the act.

These men walk a high wire fifty feet above the ground without a net.

In this act the car drives off the ramp, turns a somersault in the air, and lands upright in a net.

The Marionettes

Marionettes are puppets on strings.
The puppeteers move the strings to bring the puppets to life on stage.

This marionette show is performed two or three times a day during the fair.

Dancing...

Local groups enjoy performing at state fairs. These young people are doing traditional Ukrainian dances.

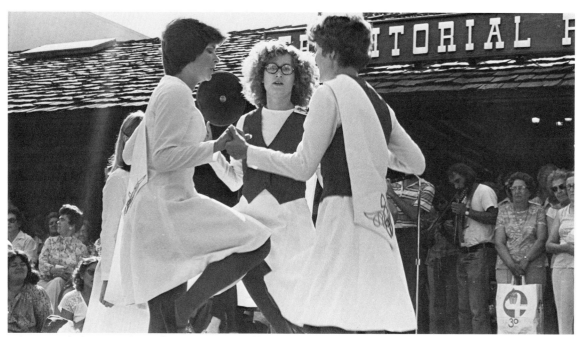

These girls are dancing an Irish jig.

& Music

Country and bluegrass music are state fair favorites. Performers come from all over the country. There are different performances about four times a day.

Baby Barnyard Animals

Baby chickens are called chicks.

Baby goats are called kids.

Baby animals are not judged. They are at the fair for children to look at.

These are young mules.

A baby pig is a squealing armful.

More Barnyard Animals

These sheep are being washed before they are judged. It's important that all the animals look their very best for the judges.

The Cow-Milking Contest

Dairy cows are milked by machines on a modern farm, but milking used to be done by hand. Each year at the fair people compete to see who can get the most milk the fastest by hand.

Cooking, Baking, & Crafts

There are hundreds of food and crafts entries exhibited each year. They fill several buildings.

Food is judged on appearance as well as taste.

This man is carving designs into leather.

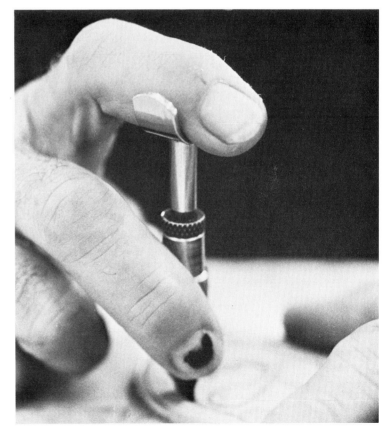

Agriculture

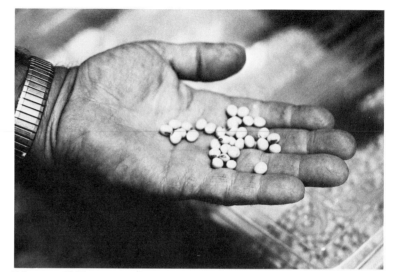

Many different kinds of crops are judged. These are soybeans.

Companies that make farm machinery exhibit their newest models at the fair.

Car Racing

Getting the car ready before the race

This driver lost control of his car by going too fast through a curve.

Drivers wear fireproof suits.

Car racing is a popular state fair event. Racers come from all over the country to compete. These cars are called Mini-Indy racers. They can go as fast as 200 miles per hour.

World Championship Rodeo

A rodeo clown

In the calf-roping contest, cowboys try to rope a calf and tie its feet faster than their opponents. On the range the calf would then be branded.

In the bulldogging contest, a cowboy jumps from his horse and wrestles a steer to the ground.

Riding a wild bucking horse can lead to falls.

The Parade

There is a parade every day at the fair.

Saint Paul Fire Department's
ORIGINAL
1905
STEAM PUMPER
PURCHASED FOR
PIONEER HOOK & LADDER
VOLUNTEERS
BY
St. Paul Fire & Marine Insurance Co.

The Midway

One of the special places at the state fair is the midway. Its games, rides, and food provide fun and entertainment for all.

The
End